Carathéodory

and

The Second Law

of

Thermodynamics

by

Jim Togeas

Winter 2018

Morris, Minnesota

ISBN 978-1-387-47506-3

Table of Contents

Foreward

The approach of Constantin Carathéodory to the Second Law of Thermodynamics is economical and elegant. Its implementation is based on an area of calculus usually not covered in the undergraduate courses in differential and integral calculus with which I am familiar. These two thoughts have shaped the writing of this monograph.

In the thirty-page section called Exposition, I have written eight short chapters that develop the mathematics and the thermodynamics of the Second Law in—this is my hope—a coherent and orderly fashion. For one not familiar with the Carathéodory approach, the coherence of the Exposition depends on the chapters being read successively.

The Carathéodory approach allows a short but rigorous derivation of the Second Law to be introduced into a course on thermodynamics. The student needs to understand the relationship between his mathematical theorem and the existence of integrating multipliers, an understanding that can be reached with simple examples provided in the Exposition. Then his statement of the Second Law allows one to introduce a universal temperature as an integrating multiplier, and the existence of entropy as a state property, which takes only a few steps, beginning at § 5 of the Exposition.

Why should the Carathéodory method challenge the older methods of Lord Kelvin (William Thomson) and Rudolf Clausius? I let the physicist Max Born respond to that question in the last section of the Exposition. It was he who encouraged Carathéodory to develop this method, and his enthusiasm for it reached such a degree that for some authors it is the Born-Carathéodory theory[*]. I'm not at all certain whether this should have been the first or last chapter in the Exposition, but finally chose the last chapter because by then the reader should know what the Carathéodory method is.

The Appendices can be read in any order or ignored if the Exposition suffices. The development of the Exposition is linear, but the Appendices are ramified, something like the way that a growing plant sends out shoots here and there in response to its environment. That reflects the growth of my own understanding as I worked on the text. The appendix on the Gibbs Paradox is not related per se to the Carathéodory theory; rather, it simply reflects my view that there is no paradox at all, but a specious one that arises from a conflation of two different experiments that are supposedly the same experiment. Not all thermodynamicists are as sanguine as Born (and me) about Carathéodory's approach, as one can read in Appendix H, but if thermodynamics is likened to a scientific sporting event, then different spectators favor different teams. That's life.

My interest in the Carathéodory method is an old one. More than fifty years ago when I was a graduate student in physical chemistry, I had familiarized my self with it from Max Born's *The Natural Philosophy of Cause and Chance*. I chose it as the topic for my graduate

---

[*] For example, Kestin, *A Course in Thermodynamics*.

student seminar, which proved to be a lucky choice for two reasons. Nobody in the audience knew anything about it, and my talk was received in a stunned silence. In other words, nobody knew enough to hoist me on my own petard. That was the other bit of luck—I was hoistable as there was much about the theory that I didn't understand.

When I retired from my teaching career, I had the leisure to return to this topic and fill in the missing gaps, which I found were more like broad canyons. As I learned, I wrote, and finally decided I'd bind my learning together in a little monograph.

Jim Togeas
Morris, Minnesota
Winter, 2018

What the hell do we care what the entropy does?

George Gamow
Mister Tompkins Explores the Atom

EXPOSITION

Carathéodory's Statement of the Second Law. In the neighborhood of any thermodynamic state, there exist states that are inaccessible by adiabatic changes whether they are reversible or irreversible.

Overview. There is a twofold aim of the Second Law:

1) to show that every thermodynamic system has a property called the entropy: $dS = dq_{rev}/T$;

2) to show that entropy in an isolated system can never decrease.

All processes compatible with the First Law of Thermodynamics can be sorted into three sets: spontaneous ($dS > 0$); equilibrium ($dS = 0$); and virtual ($dS < 0$). The third is prohibited.

Kelvin attains this twofold aim by constraining the work done by an engine operating between hot and cold reservoirs. Clausius attains it by constraining the heat flow from cold to hot reservoirs in a refrigerator. Carathéodory considers neither engines nor refrigerators, but instead examines the thermodynamic state space; in effect, he recasts the problem as one of geometry rather than engineering.

Attaining the twofold aim.

1. Pfaffian differentials. Exactness and integrability.

2. Carathéodory's Mathematical Theorem on integrating multipliers.

3. Accessibility and the ideal gas state.

4. Carathéodory's statement of the Second Law.

5. Thermodynamic temperature, θ.

6. Entropy, S.

7. Equality of thermodynamic and ideal gas temperatures.

8. Entropy and irreversibility.

9. Born's argument for preferring Carathéodory's Second Law to the statements of Kelvin and Clausius.

Appendix A. A non-integrable Pfaffian.

Appendix B. Non-uniqueness of the representation of accessible and inaccessible states.

Appendix C. Proof of the Mathematical Theorem and its sufficient condition.

Appendix D. Construction of a curve between two states accessible by an adiabatic transition.

Appendix E. On the Gibbs paradox.

Appendix F. Kelvin, Clausius, Carathéodory: A short comparison.

Appendix G. Vector representation and the necessary condition of the Mathematical Theorem.

Appendix H. A spectrum of viewpoints of the Carathéodory Second Law.

Appendix I. Biographical sketch of Constantin Carathéodory.

1. Pfaffian differentials. Exactness and Integrability.

These are equations of the form

$$dq = Xdx + Ydy + Zdz \qquad (1.1)$$

where X, Y, and Z are functions of x, y, and z. Let $\xi = \xi(x,y,z)$ be the equation of a surface for which the total differential is

$$d\xi = \frac{\partial\xi}{\partial x}dx + \frac{\partial\xi}{\partial y}dy + \frac{\partial\xi}{\partial z}dz \qquad (1.2)$$

A Pfaffian is said to be exact if all three of the conditions in Eq. (1.3) are met, and inexact otherwise.

$$\frac{\partial X}{\partial y} = \frac{\partial Y}{\partial x} \quad \& \quad \frac{\partial Z}{\partial y} = \frac{\partial Y}{\partial z} \quad \& \quad \frac{\partial X}{\partial z} = \frac{\partial Z}{\partial x} \qquad (1.3)$$

Equation (1.2) is exact by virtue of the property of mixed second-order partial derivatives that the order of differentiation is immaterial.

Exactness is required of properties of thermodynamic systems as the following argument shows. Let two three-dimensional vectors be given by

$$\vec{r} = \hat{i}x + \hat{j}y + \hat{k}z$$

$$\vec{\xi} = \hat{i}X + \hat{j}Y + \hat{k}Z$$

where the circumflex symbols denote orthonormal cartesian unit vectors. As the differential change in the first is given by

$$d\vec{r} = \hat{i}dx + \hat{j}dy + \hat{k}dz$$

3

then Eq. (1.1) for a surface can be written as

$$d\xi = \vec{\xi} \cdot d\vec{r} = Xdx + Ydy + Zdz$$

The curl of the vector is

$$\nabla \times \vec{\xi} = \hat{i}\left(\frac{\partial Z}{\partial y} - \frac{\partial Y}{\partial z}\right) + \hat{j}\left(\frac{\partial X}{\partial z} - \frac{\partial Z}{\partial x}\right) + \hat{k}\left(\frac{\partial Y}{\partial x} - \frac{\partial X}{\partial y}\right)$$

The theorem of Stokes is

$$\oint_c d\xi = \oint_c \vec{\xi} \cdot d\vec{r} = \iint_\sigma \nabla \times \vec{\xi} \cdot d\vec{\sigma}$$

The middle integral called the circulation is evaluated on a closed contour C. The integral on the right is evaluated over the surface σ that contains the contour C. Equation (1.3) is the condition that the curl vanishes, which applied to Stokes theorem gives

$$\oint_c d\xi = 0$$

But this is what is required of any thermodynamic property, that when the system undergoes a cyclical change of state the property returns to its initial value. It implies that ξ is a point function, that it has a numerical value determined purely by the thermodynamic state, and that the change in ξ

$$\Delta\xi = \xi_2 - \xi_1 = \int_{\xi_1}^{\xi_2} d\xi$$

4

is path-independent, that it is determined purely by the values of ξ in the initial and final state, and is independent of how the change occurs.

If the Pfaffian is exact or can be made exact, it is said to be *integrable*. Otherwise, it is non-integrable in the sense that it is not the differential of a point function, and that therefore the numerical value of the integral is path-dependent.

In subsequent applications of these ideas, it is useful to note that the equation for a plane tangent to a surface at x_1, y_1, and z_1 is

$$\frac{\partial \xi}{\partial x}(x - x_1) + \frac{\partial \xi}{\partial y}(y - y_1) + \frac{\partial \xi}{\partial z}(z - z_1) = 0 \tag{1.4}$$

Letting $(x-x_1) \rightarrow dx$, and similarly obtaining dy and dz, the differential expression for the tangent plane to a surface is

$$\frac{\partial \xi}{\partial x}dx + \frac{\partial \xi}{\partial y}dy + \frac{\partial \xi}{\partial z}dz = 0 \tag{1.5}$$

2. Carathéodory's Mathematical Theorem on Integrating Multipliers.

An integrating multiplier, if it exists, makes an inexact differential exact.

I paraphase the theorem as expressed by Josesph Kestin[2].

Given a Pfaffian differential expression
$$dq = X_1 dx_1 + X_2 dx_2 + X_3 dx_3 + \dots$$
where the X_i are functions of the independent variables x_i, the sufficient and necessary condition for the existence of an integrating multiplier is that in the neighborhood of any state point P there exist state points Q inaccessible for a change of state where $dq = 0$.

I explore the relationship between accessibility and integrability in bivariant and trivariant systems by working through specific instances, which will lead us to the relationship between the Mathematical Theorem and the Second Law.

Appendix C describes Carathéodory's derivation of the Theorem and makes the case that it is a sufficient condition for the existence of integrating multipliers. Appendix G makes the case that the condition is a necessary one besides

---

[2] *A Course in Thermodynamics*, p. 477. Born calls this The Theorem of Accessibility (*Natural Philosophy of Cause and Chance*, p. 144), but I have chosen to exclude the term accessibility/inaccessibility from the titles of both the Mathematical Theorem and the statement of the Second Law in order to emphasize their distinctness. The former is pure mathematics whereas the latter is an assertion about the nature of thermodynamic state space.

providing a vector test for the existence of an integrating multiplier.

The bivariant case. Consider the case

$$dq = Xdx + Ydy$$

where X and Y are functions of x and y: $X = X(x,y)$ and $Y = Y(x,y)$. Let the specific case be

$$dq = \frac{dx}{y} + \frac{dy}{x} \tag{2.1}$$

which is seen to be inexact since

$$\frac{\partial X}{\partial y} = -\frac{1}{y^2} \neq \frac{\partial Y}{\partial x} = -\frac{1}{x^2}$$

Let $dq = 0$ define a path in the x,y-plane and suppose that in the neighborhood of a point $P(x_1,y_1)$ there are points inaccessible along that path; let one such inaccessible point be $Q(x_2,y_2)$.

The question now becomes, what is the geometric structure of the plane that permits the existence of inaccessible points? The answer must be that it is filled with a family of non-intersecting curves, $\xi = \xi(x,y) = C$, where an infinity of values of C defines the space-filling curves.

Let point P lie on curve $\xi = C_1$ and Q on the adjacent curve $\xi = C_2$. These curves cannot intersect. If they did at, say, $R(x_3,y_3)$, then one could move from P to R along C_1, that is, with $d\xi = 0$, then move from R to Q along C_2, again with $d\xi = 0$, contradicting the requirement that Q is inaccessible from P along a path with dq = 0.

The condition of inaccessible points for dq = 0 must imply a family of curves and not a surface element. The geometric structure in question cannot be an area as all points within an element of area dσ would be accessible on paths for which dq = 0.

Thus, the *existence of inaccessible points implies the existence of a family of curves, which have exact Pfaffian differentials, that is, differentials that are integrable.* This in turn implies *the existence of an integrating multiplier, λ = λ(x,y)* such that λdq is exact:

$$d\xi = \lambda dq = 0 \qquad (2.2)$$

It is easy to find λ by inspection for the example posed, but let's see how it can be found analytically. By Eq. (2.1) for a bivariant system,

$$d\xi = \lambda X dx + \lambda Y dy$$

must be exact with

$$\frac{\partial \xi}{\partial x} = \lambda X \quad \& \quad \frac{\partial \xi}{\partial y} = \lambda Y$$

Taking the second cross derivatives and equating them leads to a partial differential equation for the integrating multiplier:

$$\lambda \left(\frac{\partial X}{\partial y} - \frac{\partial Y}{\partial x} \right) = Y \frac{\partial \lambda}{\partial x} - X \frac{\partial \lambda}{\partial y} \qquad (2.3)$$

For the example this becomes

$$\lambda\left(\frac{1}{x^2} - \frac{1}{y^2}\right) = \frac{1}{x}\frac{\partial\lambda}{\partial x} - \frac{1}{y}\frac{\partial\lambda}{\partial y}$$

Assuming that the integrating multiplier has the form $\lambda = x^m y^n$, the preceding equation becomes

$$\lambda\left(\frac{1}{x^2} - \frac{1}{y^2}\right) = \lambda\left(\frac{m}{x^2} - \frac{n}{y^2}\right)$$

from which it is evident that $m = n = 1$. Thus, Eq. (2.1) for the example becomes

$$d\xi = \lambda dq = xdx + ydy = 0 \qquad (2.4)$$

giving

$$\xi(x,y) = x^2 + y^2 = r^2 \qquad (2.5)$$

where I've incorporated a factor of two and the usual symbol C for the integration constant into $r^2 = 2C$.

One sees that the family of non-intersecting curves in this instance is a plane-filling family of circles of radius r centered at $x = y = 0$. By Eq. (1.5), Eq. (2.4) is just the differential form for the tangent line to a circle. For a circle of radius r, all points on the circle would be *accessible* along the curve $d\xi = 0$, but points on neighboring circles $r \pm dr$ would be *inaccessible*.

The set of circles given by Eq. (2.5) is not a unique representation of the sets of accessible and inaccessible points, which is shown in Appendix B.

In sum, I have illustrated the connection between the presence of points that are inaccessible along a curve dq = 0, the existence of exact Pfaffian differentials, and the existence of integrating multipliers for inexact differentials.

My purpose has been to illustrate a general method. However, I would be remiss in not pointing out another feature of the bivariant case, that there will always be a solution, that is, a family of curves $\xi(x,y) = C$. By contrast, higher degrees of variance will not always lead to solutions, as shown in Appendices A and G. Setting dq = 0 = Xdx + Ydy, one gets a first order differential equation,

$$\frac{dy}{dx} = -\frac{X}{Y}$$

The existence theorems of differential equations guarantee that there will always be a solution to this equation, although there may be instances where an analytical solution will not be forthcoming. For the specific case at the beginning of this section, the differential equation becomes

$$xdx + ydy = 0$$

which is the equation for a tangent line to a circle, whose equation is obtained by integration.

The trivariant case. As a specific example of the trivariant case consider

$$dq = \frac{dx}{yz} + \frac{dy}{xz} + \frac{dz}{xy}$$

(2.6)

which is inexact since

$$\frac{\partial X}{\partial y} = -\frac{1}{y^2z} \neq \frac{\partial Y}{\partial x} = -\frac{1}{x^2z}$$

One now proposes that in the vicinity of any point there are points that are inaccessible along a path dq = 0. If that is true, it follows that there must be a family of surfaces, $\xi = \xi(x,y,z) = C$ meaning that points on the surface are inaccessible to points on neighboring surfaces $\xi \pm d\xi$ along a path $d\xi = 0$. The geometric structure in question cannot be a volume as all points within an element of volume, dV would be accessible along $d\xi = 0$. The possibility that the structure is a set of curves need not be considered independently as these will simply be plane sections of the surface. The necessary existence of a family of surfaces implies the existence of an integrating multiplier that will reveal their structure.

Then the condition that

$$d\xi = \lambda dq = \lambda X dx + \lambda Y dy + \lambda Z dz \quad (2.7)$$

be exact requires in turn that

$$\frac{\partial(\lambda Y)}{\partial z} = \frac{\partial(\lambda Z)}{\partial y}$$

and two other equations given by Eq. (1.3). This leads to three partial differential equations analogous to Eq. (2.3) for the integrating multplier, one of which comes from the above equation:

$$\lambda\left(\frac{\partial Z}{\partial y} - \frac{\partial Y}{\partial z}\right) = Y\frac{\partial \lambda}{\partial z} - Z\frac{\partial \lambda}{\partial y} \qquad (2.8)$$

Assuming that

$$\lambda = x^a y^b z^c$$

and using Eqs. (1.3) and (2.8) leads to the result that one has probably already anticipated that $a = b = c = 1$. Then Eq. (2.6) becomes

$$d\xi = xdx + ydy + zdz = 0 \qquad (2.9)$$

which can be immediately integrated to give

$$\xi(x,y,z) = x^2 + y^2 + z^2 = r^2 = C \qquad (2.10)$$

Equation (2.10) of course is the equation of a sphere of radius r centered at $x = y = z = 0$, whereas Eq. (2.9) is the equation of a plane tangent to the sphere. The plane $z = 0$ projects circles of radius r,

$$x^2 + y^2 = r^2 \qquad (2.11)$$

into the x,y-plane, illustrating the notion that the case of a curve need not be considered separately.

3. Accessibility and the ideal gas state.

Here are two examples of how the First Law applied to ideal gases reveals the existence of neighboring states adjacent to the system's state that are inaccessible by reversible adiabatic transitions. The axiomatic character of the Second Law will be the proposal that inaccessibility holds for all thermodynamic systems, not just ideal gases.

Bivariant case. Given an ideal gas and the First Law of Thermodynamics, a derviation familiar from a first course in physical chemistry shows that in an adiabatic reversible change of state

$$\xi(p, V) = pV^{\gamma} = C$$

(3.1)

where the adiabatic parameter is $\gamma = C_p/C_v$. A choice of pressure and volume defines the system's thermodynamic state. The indicator diagram (p,V-plane) is filled with a family of non-intersecting curves given by Eq. (3.1), each defined by a different value of C. For a reversible adiabatic change of state, $d\xi = 0$: all points lying on the curve $\xi = C$ are accessible by a reversible, adiabatic change, but those on adjacent curves are inaccessible.

Trivariant case. A closed system consists of an ideal gas divided into two compartments at the same temperature but with variable volumes; the conditions in the two compartments are

$$\{n_1, V_1, p_1, T\} \ \& \ \{n_2, V_2, p_2, T\}$$

Now change the volumes reversibly by letting the external pressure on V_1 be p_1, and on V_2 be p_2. The First Law for a change of state is then[3]

$$dU = C_v dT = dq - p_1 dV_1 - p_2 dV_2$$

where C_v is the sum of the constant volume heat capacities for the two compartments. We now show that all points accessible adiabatically fall on a surface. Setting $dq = 0$, rearranging, using the ideal gas law, and dividing through by the temperature gives

$$C_v \frac{dT}{T} + n_1 R \frac{dV_1}{V_1} + n_2 R \frac{dV_2}{V_2} = 0$$

Dividing through this expression by nR, where n is the total number of moles of gas, letting x represent the mole fraction, and setting $C_v / nR = c$, gives

$$c \frac{dT}{T} + x_1 \frac{dV_1}{V_1} + x_2 \frac{dV_2}{V_2} = 0$$

The expression has the form

$$XdT + YdV_1 + ZdV_2 = 0$$

and is exact. Taking the indefinite integral gives

$$\xi(T, V_1, V_2) = T^c V_1^{x_1} V_2^{x_2} = C$$

$$(3.2)$$

---

[3] The internal energy of an ideal gas depends only on its temperature: $U = U(T)$. This conclusion depends only on the First Law and the Joule-Thomson experiment. The point here is that the argument is not circular, that is, that the Second Law is not hidden within the argument.

14

where C is a constant as usual.

The geometric structure of this surface is easy to understand. Consider the following Pfaffian, which is exact:

$$d\xi = yzdx + xzdy + xydz$$
$$d\xi = d(xyz)$$

meaning that

$$\xi = xyz = C,$$

which is the formula for an equilateral hyperboloid. The state space, T, V_1, V_2 is filled with equilateral hyperboloids obeying Eq. (3.2) with each surface characterized by a different value of C. The accessible points lie on a surface of constant C, while the inaccessible neighboring points lie above and below on neighboring hyperboloids characterized by $C \pm dC$.

4. Carathéodory's statement of the second law.

In the neighborhood of any thermodynamic state there exist states that are inaccessible by adiabatic changes whether they are reversible or irreversible.

Carathéodory writes the First Law[4] as

$$\Delta U - w = U_f - U_i - w = 0$$

$$(4.1)$$

---

[4] Chemist's convention: dU = dq+dw. Born, *Natural Philosophy*, p. 36-38; Georgiadou, §2.2.

Heat is that quantity that causes Eq. (4.1) to be non-vanishing. This is rather at odds with the approach in many texts on thermodynamics where one begins with Joule's experiments on establishing the mechanical equivalent of heat. Carathéodory and Born say that to replace zero by q, heat, is to define heat in mechanical terms.

If there are inaccessible states, then there is an integrating multiplier for the Pfaffian differential. The examples in § 3 show that such states exist in bivariant and trivariant systems of an ideal gas, but do they exist in every thermodynamic system, in the liquid, solid, and solutions states of arbitrary variance?

Just as with the First Law, the Second Law is taken to be axiomatic—plausible, but axiomatic. The Kelvin and Clausius statements are axiomatic in prohibiting certain transformations of heat. Carathéodory's statement likewise has its prohibition, viz., that some states neighboring a given state cannot be reached by an adiabatic process. They are negative statements, statements of impossibility, and of course the impossible can never be demonstrated with absolute rigor. Instead it is the fruitfulness of the axiom that gives us confidence in its validity.

The three statements of the second law are equivalent in the sense that if one is true all three must be true, and conversely if one is false they must all be false. Demonstrations of equivalence are in the literature.[5] I compare the three in Appendix F.

5. Thermodynamic temperature, θ.

The integrating multiplier generates a thermodynamic property from the heat flow:

$$d\xi = \lambda dq \qquad (5.1)$$

The analysis below argues that the multiplier is a universal function of an empirical temperature scale.

Consider a trivariant thermodynamic system,

$$\xi = \xi(V_1, V_2, t) \qquad (5.2)$$

where t is any empirical temperature scale. The system consists of two compartments of variable volume at the same temperature t, but their content is no longer constrained to an ideal gas as in § 3. Heat can be exchanged between the compartments:

---

[5] Fermi demonstrates the equivalence of the Kelvin and Clausius statements in § 7 of his wonderful book. Chandrasekhar in his book on stellar structures states that the demonstrations with the Carathéodory statement "involve some rather delicate considerations" (p. 34) that go beyond the scope of his book and refers the interested reader to T. Ehrenfest Afanassjewa, *Zs. f. Phys.* **1925**, *33*, 933.

$$dq = dq_1 + dq_2 \qquad (5.3)$$

As we assume the existence of inaccessible points, there must be an integrating multiplier for each compartment as well as for the entire system, so Eq. (5.3) becomes

$$\frac{d\xi}{\lambda} = \frac{d\xi_1}{\lambda_1} + \frac{d\xi_2}{\lambda_2} \qquad (5.4)$$

Naturally,

$$\xi_1 = \xi_1\left(V_1, t\right), \lambda_1 = \lambda_1\left(V_1, t\right)$$

$$\xi_2 = \xi_2\left(V_2, t\right), \lambda_2 = \lambda_2\left(V_2, t\right)$$

Now change the independent variables in Eq. (5.2) to

$$\xi = \xi(\xi_1, \xi_2, t) \qquad (5.5)$$

with a total differential of

$$d\xi = \frac{\partial \xi}{\partial \xi_1} d\xi_1 + \frac{\partial \xi}{\partial \xi_2} d\xi_2 + \frac{\partial \xi}{\partial t} dt \qquad (5.6)$$

Comparing Eqs. (5.4) and (5.6) gives

$$\frac{\partial \xi}{\partial \xi_1} = \frac{\lambda}{\lambda_1}, \frac{\partial \xi}{\partial \xi_2} = \frac{\lambda}{\lambda_2}, \,\& \, \frac{\partial \xi}{\partial t} = 0 \qquad (5.7)$$

Now it follows that

$$\frac{\partial}{\partial \xi_1}\left(\frac{\partial \xi}{\partial t}\right) = 0 = \frac{\partial}{\partial t}\left(\frac{\lambda}{\lambda_1}\right)$$

$$\frac{\partial}{\partial \xi_2}\left(\frac{\partial \xi}{\partial t}\right) = 0 = \frac{\partial}{\partial t}\left(\frac{\lambda}{\lambda_2}\right) \qquad (5.8)$$

As the right-hand side of Eq.(5.8) vanishes, we find

$$\frac{1}{\lambda}\frac{\partial\lambda}{\partial t} = \frac{1}{\lambda_1}\frac{\partial\lambda_1}{\partial t} = \frac{1}{\lambda_2}\frac{\partial\lambda_2}{\partial t}$$

(5.9)

As λ_1 depends on V_1 and t, and λ_2 depends on V_2 and t, and as the two volumes can be chosen arbitrarily, it follows that the second equality can hold only if there is a universal function of the temperature, t, $f = f(t)$. Then by Eq. (5.9),

$$\frac{\partial\ln\lambda}{\partial t} = -f(t)$$

(5.10)

with the minus sign chosen for convenience. Integration gives

$$\ln\lambda = -\int f(t)dt - \ln A$$

(5.11a)

where A can be any function of ξ_1 and ξ_2. Then

$$\lambda = \frac{1}{A}\exp\left(-\int f(t)dt\right)$$

$$= \frac{C}{A(\xi_1,\xi_2)C\exp\left(\int f(t)dt\right)}$$

(5.11b)

Now define the thermodynamic temperature to be

$$\theta = C\exp\left(\int f(t)dt\right)$$

(5.12)

The constant C supples units to θ; by choosing $C \geq 0$, then $\theta \geq 0$—the thermodynamic temperature will never be a negative quantity. The functionality defined by Eq. (5.5) means that the integrating multiplier as given in Eq. (5.11b) has factored into a temperature and a term dependent on the nonthermal variables ξ_1 and ξ_2:

$$\lambda = \frac{C}{A\left(\xi_1, \xi_2\right)\theta}$$

(5.13).

6. Entropy, S.

The entropy of a system is defined using Eqs. (5.1) and (5.13),

$$dS = \frac{dq_{rev}}{\theta} = \frac{A\left(\xi_1, \xi_2\right)}{C}d\xi$$

(6.1a)

giving on indefinite integration

$$S = \frac{1}{C}\int A\left(\xi_1, \xi_2\right)d\xi + S_0$$

(6.1b)

Similarly, the entropies of the subsystems can be defined by

$$dS_k = \frac{dq_{k,rev}}{\theta} = \frac{A_k\left(\xi_k\right)}{C}d\xi_k$$

(6.2a)

and

$$S_k = \frac{1}{C}\int A_k\left(\xi_k\right)d\xi_k + S_{k,0}$$

(6.2b)

where $k = 1,2$. This establishes that entropies are additive:

$$S = S_1 + S_2$$

(6.3a)

and

$$S_0 = S_{1,0} + S_{2,0}$$

(6.3b)

7. Equality of thermodynamic and ideal gas temperatures.

The argument depends on the fact that the energy of an ideal gas depends only on the temperature and not on mechanical variables such as the volume. A nice argument can be made from the fundamental equation, $dU = TdS - pdV$, and the Maxwell relations, but this presupposes that we have already established that $\theta = T$. However, the Joule-Thomson experiment leads to the same conclusion, that $U = U(T)$, while depending only on the First Law.

In this experiment, a gas is subjected to a sudden (irreversible) pressure drop and expansion, and a change in temperature, if any, is measured. The experiment occurs at constant enthalpy:
$$H_i = U_i + p_i V_i = H_f = U_f + p_f V_f$$
where i and f refer to the initial and final states. In the ideal gas limit this becomes
$$H_i = U_i + nRT_i = H_f = U_f + nRT_f$$
In this same limit, experiment shows that the temperature does not change; hence
$$U_i = U_f$$
As the volume has changed but the internal energy has not, it follows that the internal energy of an ideal gas does not depend on its volume but only

on its temperature. Thus, for a reversible change of state in an ideal gas the First Law becomes

$$dU = C_v dT = dq_{rev} - p_1 dV_1 - p_2 dV_2 \quad (7.1)$$

If the change of state is also isothermal, then setting $dT = 0$, rearranging, and using the ideal gas law, one has

$$\frac{dq_{rev}}{T} = R\left(n_1 \frac{dV_1}{V_1} + n_2 \frac{dV_2}{V_2}\right) \quad (7.2)$$

Note how T has been partitioned from the non-thermal state variables as in the generic equations (5.13) and (6.1):

$$\frac{dq_{rev}}{\theta} = A\left(\xi_1, \xi_2\right) d\xi_1 d\xi_2 \quad (7.3)$$

Hence,

$$\theta = T \quad (7.4)$$

As the thermodynamic temperature is a universal function, establishing the equality for one case establishes it for all cases.

Thus,

$$dS = \frac{dq_{rev}}{T} \quad (7.5a)$$

It follows that

$$\Delta S = \int_i^f dq_{rev}/T \quad (7.5b)$$

is path-independent, depending only on the choice of initial and final states, and is the same for that choice whether the process is reversible or irreversible.

8. Entropy and irreversibility.

Entropy is defined by Eq. (7.5a). For a reversible, adiabatic change of state, $\Delta S = 0$. It is clear, however, given the definition of entropy, that if the process is irreversible

$$\Delta S \neq \int_i^f dq_{irr}/T$$

$$(8.1)$$

Consider an irreversible change of state in the trivariant case of § 3, but now the system need not be an ideal gas. The independent variables are (V_1, V_2, S), where S is the total entropy of the system, which has adiabatic walls so that thermal exchanges with the surroundings need not be considered. The entropy change is

$$\Delta S = S_f - S_i \neq 0 \qquad (8.2)$$

even though

$$\int_i^f dq_{irr}/T = 0$$

$$(8.3)$$

Here is a striking difference between reversible and irreversible behavior: if the process is adiabatic, dq $= 0$ in both cases, but $\Delta S = 0$ in the former case but $\Delta S \neq 0$ in the latter.

In what follows, the point will be to show that in an irreversible adiabatic process the entropy must increase:

$$\Delta S > \int_i^f dq_{irr}/T = 0$$

The procedure will be to illustrate the argument with an ideal gas, make a general argument, and illustrate that with another ideal gas calculation.[6]

8.1. A simple argument. The standard entropy of neon gas (1 bar = 0.9869 atm) and 298.15 K is 146.33 J mol^{-1} K^{-1}. I calculate the entropy of two neighboring states at the same temperature. This does not presuppose that entropy must increase or decrease in an adiabatic irreversible process.

Three neighboring states at 298.15 K[7]

State	p(atm)	V(L)	S	C
1	0.9769	25.04	146.41	209.37
2	0.9869	24.79	146.33	208.00
3	0.9969	24.54	146.25	206.59

Each state lies along an adiabat $pV^{5/3} = C$, on which entropy is a constant. Hence, states along

_____

[6] For the ideal gas illustrations, I use numerical arguments. The general case is the argument of Chandrasekar, who writes that he followed review articles by Born and Landé.

[7] Units in columns 4 and 5 are J/(mol-K) and atm-L$^{5/3}$.

adiabat 2, for example, are accessible by a *reversible* adiabatic change of state, but those on adiabats 1 and 3 inaccessible from 2 reversibly. Let state 2 be the initial state for an *irreversible* adiabatic change of state. Then let the final state be either adiabat 1 or adiabat 3. It must be one or the other: if both could be achieved, then there would be no states in the vicinity of adiabat 2 that were inaccessible by an adiabatic transition in violation of Carathéodory's statement of the second law.

8.2 Appeal to experiment. For any change of state, reversible or irreversible,

$$dU = TdS - pdV \qquad (8.4)$$

which Gibbs called the fundamental equation. Then

$$dS = \frac{dU}{T} + \frac{p}{T}dV$$

Let an ideal gas expand irreversibly in a Joule-Kelvin apparatus, a process that occurs isothermally and adiabatically. As $pdV/T = nRdV/V = -nRdp/p$, we have

$$\Delta S = nR\ln\left(\frac{V_f}{V_i}\right) = -nR\ln\left(\frac{p_f}{p_i}\right) > 0 \qquad (8.5)$$

Hence, state 1 is accessible from state 2, but state 3 is inaccessible—the entropy increases in this adiabatic, irreversible process.

8.3. The argument of Chandrasekhar. In this more general argument, it is not assumed that the system is an ideal gas. Initially, the system is in the state (V_{1i}, V_{2i}, S_i) and finally in (V_{1f}, V_{2f}, S_f). Go to an intermediate state with the final volumes but the same entropy by an adiabatic, *reversible* process:

$$(V_{1i}, V_{2i}, S_i) \rightarrow (V_{1f}, V_{2f}, S_i)$$

Now go to the final state by an isochoric, adiabatic, *irreversible* process such as frictional rubbing:

$$(V_{1f}, V_{2f}, S_i) \rightarrow (V_{1f}, V_{2f}, S_f)$$

If in some instances it is possible that the final entropy is less than the initial and in others that it is greater, then there will no states neighboring the initial state that are inaccessible adiabatically. Hence, entropy must always increase in an adiabatic, irreversible process or it must always decrease. Appeal to experiment shows that it must always increase.

8.4. Surfaces of constant entropy in an ideal gas.

For the trivariant ideal gas system in § 3,

$$dU = C_v dT = dq - p_1 dV_1 - p_2 dV_2$$

Using the definition of the entropy in Eq. (7.5a) and the ideal gas law, this becomes

$$dS = C_v \frac{dT}{T} + n_1 R \frac{dV_1}{V_1} + n_2 R \frac{dV_2}{V_2} \qquad (8.6)$$

This equation can be used to calculate the entropy change for both reversible and irreversible processes because changes in system properties are path-independent, and depend only on the choice of intial and final states.

An isentropic change occurs adiabatically and reversibily, $dS = 0$. In § 3 one sees that this leads to the equation for an equilateral hyperboloid:

$$\xi\left(T, V_1, V_2\right) = T^c V_1^{x_1} V_2^{x_2} = C$$

(3.2)

Hence, one associates the integration constant with a definite value of the entropy:

$$C \Leftrightarrow S$$

These thoughts provide a point of departure for both § 8.5 and for resolving the so-called Gibbs paradox purely in terms of classical thermodynamics, which is the subject of Appendix E.

8.5 Illustrating Chandrasekhar's argument with an ideal gas calculation.

As stated in § 8.1, a mole of neon gas at 298.15 K and in its standard state has an entropy of 146.33 J mol^{-1} K^{-1} and a volume of 24.79 L. For a monatomic ideal gas such as neon, the constant $c = 3/2$ in Eq. (3.2).

Let the initial state in Chandrasekhar's argument be

$$\{T = 298.15 \text{ K}, V_1 = V_2 = 12.39_5 \text{ L}\}$$

The mole fractions will be

$$x_k = V_k / (V_1 + V_2) = \tfrac{1}{2}$$

for both k = 1,2. Then

$$C = 6.3626 \times 10^4 \, K^{3/2} \, L$$

a number associated with the value of the entropy stated above.

Now, following Chandrasekhar's argument, change the volumes reversibly and adiabatically to

$$\{V_1 = 10.00 \text{ L}, \; V_2 = 14.79 \text{ L}\}$$

with final mole fractions of 0.4034 and 0.5966, resp. This change occurs isentropically, that is, it simply moves the state point to a different location on the hyperoloid surface. The new temperature will be

$$T = \left[\frac{C}{V_1^{x_1} V_2^{x_2}} \right]^{2/3} = 293.87 \, K$$

Once again following Chandrasekhar's argument, raise the temperature irreversibly and isochorically by frictional rubbing or mechanical stirring. Let the new temperature be T* = 300 K. Then by Eq. (8.6), the entropy will be

$$S = 146.33 + \frac{3}{2} nR \ln \left(\frac{T^*}{T} \right) = 146.59 \, J \, mol^{-1} K^{-1}$$

with n = 1. Thus, the entropy increases in the irreversible step, which was the point to be illustrated.

9. Born's argument for preferring Carathéodory's Second Law to the statements of Kelvin and Clausius.

Max Born[8] reflects on how thermodynamics developed from the theory of heat engines and statements of Kelvin and Clausius about transformations of heat that are impossible. While expressing his admiration for their achievement, he writes that their methods

> deviated too much from the ordinary methods of physics; I discussed the problem with my mathematical friend, Carathéodory, with the result that he analyzed it and produced a much more satisfactory solution.

Born further reflects that

> [t]he principles from which Kelvin and Clausius derived the second law are formulated in such a way as to cover the greatest possible range of processes incapable of execution...Carathéodory remarked that it

---

[8] Max Born (1882-1970) was a mathematical physicist who with Werner Heisenberg and Pascual Jordan established the matrix mechanics representation of quantum mechanics. He was awarded the Nobel Prize in physics in 1954. He held a chair in theoretical physics at Göttingen until 1933 when the Nazis forced his resignation because he was Jewish. He took a position in theoretical physics at Edinburgh in 1936.

is perfectly sufficient to know the existence of *some* impossible processes to derive the second law. I need hardly say that this is a logical advantage.

For Born, the deviation "from the ordinary methods of physics" meant that the mathematical underpinnings of the Kelvin and Clausius statements were insufficient. I understand this to mean that whereas, for example, quantum theory is underpinned by the commutation relations of operators and the theory of vector spaces, and the kinetic molecular theory by the methods of probability, thermodynamics had no analogous mathematical foundation. Carathéodory overcame this shortcoming by deriving the The Mathematical Theorem, developing the Pfaffian calculus with its integrating multipliers, and relating these to the notions of accessibility.[9]

---

[9] The two quotes are from *Natural Philosophy*, pp. 38-39. But also see Georgiadou, § 2.2 for additional thoughts by Born.

Appendix A. A non-integrable Pfaffian.

The Pfaffian

$$dq = ydx - xdy + zdz$$

is non-integrable. The proof is by contradiction.

Assume that an integrating multiplier, λ, exists such that $d\xi = \lambda dq$ is exact. Then the conditions for exactness such as

$$\frac{\partial}{\partial y}\left(\frac{\partial \xi}{\partial x}\right) = \frac{\partial}{\partial x}\left(\frac{\partial \xi}{\partial y}\right)$$

lead to the following equations for the integrating multiplier:

$$x\frac{\partial \lambda}{\partial x} + y\frac{\partial \lambda}{\partial y} = -2\lambda \tag{A.1}$$

and

$$\frac{\partial \lambda}{\partial x} = \frac{y}{z}\frac{\partial \lambda}{\partial z} \quad \& \quad \frac{\partial \lambda}{\partial y} = -\frac{x}{z}\frac{\partial \lambda}{\partial z} \tag{A.2}$$

Substituting Eqs. (A.2) into (A.1) gives $\lambda = 0$, showing that the assumption of an integrating multiplier is incorrect.

Appendix B. Non-uniquenss of the representation of accessible and inaccessible points.

Let Eq. (1.1) be inexact, and let there be neighboring points that are inaccessible from a point along a path where dq = 0. Then by the inaccessibility principle, an integrating multiplier, λ, exists such that dξ = λdq is exact. One must have

$$\lambda = \frac{d\xi}{dq} = \frac{\partial\xi/\partial x}{X} = \frac{\partial\xi/\partial y}{Y} = \frac{\partial\xi/\partial z}{Z} \qquad (B.1)$$

The following procedure leads to different forms of the integrating multiplier for the inexact differential Eq. (2.1). Let

$$x = \sqrt{u} \quad \& \quad y = \sqrt{v} \qquad (B.2)$$

Then

$$dq = \frac{du + dv}{2\sqrt{uv}} \qquad (B.3)$$

Evidently,

$$\lambda = 2\sqrt{uv}$$

giving dξ = du + dv, and

$$\xi = u + v = C \qquad (B.4)$$

In the analysis in § 2, the locus of accessible points for a given point was a circle of radius r centered on x = y = 0, and the loci of neighboring inaccessible points were on circles of radii r ± dr. Equation (B.4) defines a straight line of slope -1 in the u,v-plane. The locus of accessible points for a given point is a straight line of constant C, and the loci of neighboring inaccessible points are on straight lines $C \pm dC$. Hence, instead of a mapping of concentric circles we get a mapping of parallel straight lines.

A different mapping is obtained by assuming that the integrating multiplier has the form

$$\lambda = \exp\left[\alpha(x,y)\right]$$

(B.5)

where α is a function to be determined. Then the equivalent of Eq. (2.3) leads to the following differential equation:

$$\frac{1}{y}\frac{d\alpha}{dy} - \frac{1}{y^2} = \frac{1}{x}\frac{d\alpha}{dx} - \frac{1}{x^2} = -\beta$$

(B.6)

where β is an arbitrary "separation constant" assumed to be positive. This is easily integrated to give α, in which case the integrating multiplier becomes

$$\lambda = xy \, \exp\left[-\frac{\beta}{2}\left(x^2 + y^2\right)\right]$$

(B.7)

from which it seen that

$$d\xi = \exp\left[-\frac{\beta}{2}\left(x^2 + y^2\right)\right]\left(xdx + ydy\right)$$

(B.8)

is exact.

A simple substitution into Eq. (B.8) leads to yet another form of the exact differential. Setting

$$u = \frac{\beta}{2}x^2 \quad \& \quad v = \frac{\beta}{2}y^2$$

(B.9)

leads to

$$d\xi = \frac{1}{\beta}\exp\left[-(u+v)\right]dudv$$

(B.10)

which, aside from constants, is the transformation Eq. (B.2). On integration, and incorporation of all constants into the constant C, one has

$$\xi = \exp\left[-(u+v)\right] = C$$

(B.11)

Equation (B.11) is the solution of Eq. (B.8), which returns us to both Eqs. (2.5) and (B.4):

$$x^2 + y^2 = r^2 = -\ln(\xi)$$
$$u = -v - \ln(\xi)$$

Here is one more example. Use plane polar coordinates, $x = r\cos\theta$ and $y = r\sin\theta$, in Eq. (B.8) to obtain

$$d\xi = \exp\left[-\frac{\beta r^2}{2}\right]rdr = 0$$

This is readily integrated using the substitution $u = \beta r^2/2$ with the limits $0 \le u \le \infty$, to give

$$\xi = \frac{1}{\beta} = C$$

(B.12)

Accessible points with $dq = 0$ lie on a horizontal line with neighboring inaccessible points on neighboring lines of $\beta \pm d\beta$.

Thermodynamic constraints will limit the distribution of accessible/inaccessible points.

Appendix C. Proof of Carathéodory's Mathematical Theorem.

C.1. Carathéodory/Born.[10] Given an inexact Pfaffian,

$$dq = Xdx + Ydy + Zdz = 0 \qquad (C.1)$$

and given that from a state point there are neighboring points adiabatically inaccessible, the goal is to prove that there must exist a surface of accessible points, and hence that an integrating multiplier exists. Consider the following diagram:

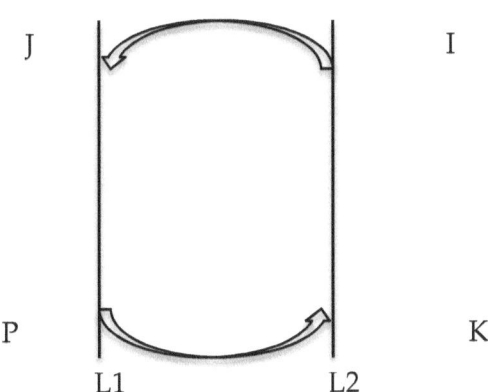

i) Point I is inaccessible adiabatically from P.

ii) Construct a straight line L1 through P that does not intersect I. L1 will not be a solution curve of Eq. (C.1).

---

[10] Born, *Natural Philosophy*, App. 7. Carathéodory in Kestin, *Second Law*, pp. 240-241.

iii) Beginning at I, construct a curve $\phi(x,y,z) = C$ that satisfies (C.1) and intersects L1. Appendix D shows how this can be done. Label the point of intersection J. As $dq \neq 0$ along L1, condition i) will not be violated.

iv) Construct a straight line L2 through point I and parallel to L1. As in ii), $dq \neq 0$.

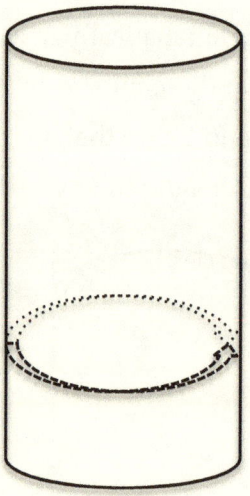

v) Choose a point K on L2 and beginning at point P construct a curve $\theta(x,y,z) = C'$ to it such that $dq = 0$.

vi) Now roll the plane defined by L1 and L2 into a cylinder such that L1 and L2 coincide with P falling on K. This desideratum dictated the choice of K in the preceding step. The cylinder can always be formed without tearing the plane or changing the distance between points. As a result the curve from PKP, $\theta = C'$, will be a closed curve on which dq = 0.

Result: There is a closed curve PKP along which dq = 0, with neighboring points J and I that cannot be reached adiabatically from the curve. If J doesn't fall on I, that is of no consequence.

<u>Conclusion</u>. The curve "can be made, by steady deformation of the cylinder, to describe a surface which contains obviously all solutions starting from P."[11] However, we chose P arbitrarily: "by varying its position we obtain a family of surfaces, $\xi = \xi(x,y,x) = C$, that depend on the parameter C and that contains all curves of" dq = 0.[12] As

$$d\xi = \frac{\partial \xi}{\partial x}dx + \frac{\partial \xi}{\partial y}dy + \frac{\partial \xi}{\partial z}dz = 0$$

then dq and dξ must be proportional, which implies the existence of an integrating multiplier:

$$d\xi = \lambda(x,y,z)dq = 0$$

<u>A sufficient condition</u>. One sees by this argument that *the existence of inaccessible points in the vicinity of any state point is a sufficient condition for the existence of an integrating multiplier.*

Kestin remarks that Carathéodory's proof and an analogous one by Born "are to a certain extent intuitive and cannot be regarded as entirely rigorous. A more abstract, analytic proof was given

---

[11] Born, p. 146.
[12] Carathéodory, p. 241.

by H. A. Buchdahl."[13] In this same section, Kestin provides a Carathéodory-type geometric proof that he regards as more rigorous. I discuss Buchdahl's work in the next section.

C.2. Buchdahl[14]. He offers an alternative proof of the principle, which he formulates as follows:

> **In the neighborhood of any arbitrary point G_0 there are points G which are not accessible from G_0 along solution curves of the equation**
>
> $$dq = \vec{R} \cdot d\vec{r} = Xdx + Ydy + Zdz = 0$$
>
> **if, and only if, the equation is integrable.**

Integrability means, of course, either that

$$\nabla \times \vec{R} = 0$$

or that there exists an integrating multiplier

$$d\xi = \lambda dq$$

such that

$$\nabla \times \vec{\xi} = 0$$

Note that Buchdahl is working the inverse of the Born/Carathéodory problem, that the latter moves from inaccessibility to integrability whereas the former moves from integrability to inaccessibility. In what follows I discuss Buchdahl's argument in some detail, but omit the laborious part of his proof.

---

[13] Kestin, *Course in Thermodynamics*, p. 478.

[14] "On the Theorem of Carathéodory." I've changed his notation to be consistent with mine.

Note that Buchdahl assumes that the Pfaffian is integrable; for convenience I will assume that dq is inexact, which, given his assumption, means that an integrating multiplier exists. As usual let $\xi = \xi(x,y,z)$ and set[15]

$$X = \frac{\partial \xi}{\partial x} \quad \& \quad Y = \frac{\partial \xi}{\partial y}$$

(C.2)

and

$$Z = \frac{\partial \xi}{\partial z} + P(x,y,z)$$

(C.3)

which defines $P(x,y,z)$. Substituting Eqs. (C.2) and (C.3) into (C.1) gives

$$dq = \frac{\partial \xi}{\partial x} dx + \frac{\partial \xi}{\partial y} dy + \frac{\partial \xi}{\partial z} dz + P(x,y,z) dz = 0$$

which is the same as

$$\frac{d\xi}{dz} = -P(x,y,z)$$

(C.4a)

If Eq. (C.4a) can be integrated to give

$$\xi = \xi(x,y,z) = C$$

(C.5)

then that is equivalent to finding an integrating multiplier. But then the question becomes, can it be integrated?

At this point, Buchdahl makes the substitution

$$P(x,y,z) \rightarrow P(\xi,y,z)$$

likening it to a Legendre transformation in thermodynamics where one, for example, changes

---

[15] I've modified the definitions in Eqs. (C.2) and (C.3) slightly.

the mechanical variable from pressure to volume. Hence the equation to solve becomes

$$\frac{d\xi}{dz} = -P(\xi, y, z)$$

(C.4b)

By hypothesis Eq. (C.4b) is integrable. By a laborious argument that I will not reproduce, he shows that this can be true only if

$$\frac{\partial P}{\partial y} = 0$$

(C.6)

Then the differential equation is

$$\frac{d\xi}{dz} = -P(\xi, z)$$

(C.4c)

This is an ordinary first-order differential equation of one dependent and one independent variable, which always has a solution. The consequence of $P(\xi, z)$ no longer depending *explicitly* on the variable y is that in the vicinity of a state point G_0 there are state points G that are inaccessible on the solution curves of Eq. (C.4c). Of course, that is the point to be proved, that integrability implies inaccessibility.

C.3. Comment on these two proofs of Carathéodory's Mathematical Theorem. I've noted that Kestin remarks that the Born/Carathéodory proof is intuitive and not as rigorous as he'd like. I note that the complete development of Buchdahl's ideas in the last paragraph of § C.2 is laborious. However, in my view it is not necessary to follow Buchdahl in all that detail as the vector argument in Appendix G is rigorous and not laborious, and is equivalent to Buchdahl's as footnote 3 in his paper indicates.

Appendix D. Constructing a curve between two states accessible via an adiabatic process.

D1. The general case.

Given a Pfaffian expression,

$$dq = Xdx + Ydy + Zdz$$

reduce the number of independent variables from three to two via a transformation, $x = x(u,v)$, $y = y(u,v)$, and $z = z(u,v)$, as we want two variables to construct a curve. As

$$dx = \frac{\partial x}{\partial u}du + \frac{\partial x}{\partial v}dv$$

(D.1)

with similar total differentials for dy and dz, then the Pfaffian may be rewritten

$$dq = Udu + Vdv \qquad (D.2)$$

where

$$U = X\frac{\partial x}{\partial u} + Y\frac{\partial y}{\partial u} + Z\frac{\partial z}{\partial u}$$

(D.3)

and

$$V = X\frac{\partial x}{\partial v} + Y\frac{\partial y}{\partial v} + Z\frac{\partial z}{\partial v}$$

(D.4)

By setting $dq = 0$, Eq. (D.2) becomes a first-order differential equation

$$\frac{du}{dv} = -\frac{V}{U}$$

(D.5)

that can always be integrated to give a curve $\phi(u,v) = C$ along which $dq = 0$.

D.2. An example. The example is the inexact Pfaffian given by Eq. (2.6):

$$dq = \frac{dx}{yz} + \frac{dy}{xz} + \frac{dz}{xy}$$

If, for example, (x,y,z) are units of length, and if (u,v) likewise are units of length, a simple transformation compatible with the units is

$$x = u, y = v, z = \sqrt{uv}$$

After a bit of calculating

$$U = \frac{2u+v}{2(uv)^{3/2}} \quad \& \quad V = \frac{2v+u}{2(uv)^{3/2}}$$

Although Eq. (2.6) is inexact, Eq. (D.5) is exact because the mutual denominator in U and V cancels when one constructs the ratio, V/U.

However, there's a deeper significance in that the denominator is the integrating multiplier:

$$\lambda(x,y,z) = \lambda(u,v) = 2xyz = 2(uv)^{3/2}$$

Then one has

$$d\phi = \lambda dq = 2udu + 2vdv + d(uv) = 0$$

which gives

$$\phi(u,v) = u^2 + v^2 + uv = C$$

However, if we now transform back to three dimensions, we find

$$\xi(x,y,x) = x^2 + y^2 + z^2 = r^2$$

as usual identifying the constant with the square of the radius of the sphere. Transforming to plane polar coordinates, $u = \rho\cos\phi$ and $v = \rho\sin\phi$, and choosing the sphere's radius equal to unity gives

$$\rho^2 = \frac{2}{2 + \sin(2\phi)}$$

A polar graph shows a closed curve of bilateral symmetry[16] as expected from the generic curve on the cylinder's surface—see p. 36. Inspection of the polar formula shows that

$\rho = 1$ for $\phi = 0, \pi/2, \pi, 3\pi/2, 2\pi$

$\rho(\text{max}) = \sqrt{2}$ for $\phi = 3\pi/4, 7\pi/4$

$\rho(\text{min}) = \sqrt{2/3}$ for $\phi = \pi/4, 5\pi/4$

The area enclosed by the curve might be of interest if it had a significant relationship to the area of a unit sphere, 4π, but since the area is bounded by

$$2\pi/3 < \text{area} < 2\pi$$

it seems uninteresting[17]; the upper bound of course is the sphere's circumference.

One way to look at this result is that

$$\xi(x,y,z) \rightarrow \phi(u,v)$$

is just the mapping of a sphere onto a plane. But more to the point, it illustrates the remark of Born in Appendix C.1, that

$$\phi(u,v) \rightarrow \xi(x,y,z)$$

is the "steady deformation of (the curve on) the cylinder to describe a surface which contains obviously all solutions starting from P." However, my imagination's eye fails to see how this steady

---

[16] The closed curve belongs to the D_{2h} point group.
[17] By integrating to get an equation in closed form, I find the area to be 2.1535, close to the lower limit.

deformation would work, but the transformation equations "see it" well enough.

Appendix E. On the Gibbs paradox.

E.1. Ideal gas mixing. The paradox begins with the formula for ideal gas mixing, aka diffusion. Let n_1 and n_2 moles of different ideal gases be separated by a partition in an apparatus with adiabatic walls. When the partition is removed, the gases mix irreversibly. Hence we have an irreversible, adiabatic process that results in an entropy increase. The gas mixing formula is obtained as follows.

The initial conditions with the partition in place are $\{T, p, n_1, V_1\}$ and $\{T, p, n_2, V_2\}$. A crucial observation is that the two compartments are at the same pressure. When the partition is removed, the two gases mix isothermally and isoergically, which is a property of ideal gases. The final condition is

$$\{T, p = p_1 + p_2, n = n_1 + n_2, V = V_1 + V_2\}$$

Overall the gas mixing occurs isobarically, the sum of the partial final pressures adding up to the initial pressures in each compartment, a consequence that follows from Boyle's Law.

The fundamental equation is

$$dU = TdS - pdV$$

With $dU = 0$, the change in the total entropy is given by

$$dS = dS_1 + dS_2$$

where

$$dS_1 = \frac{pdV_1}{T} = n_1 R \frac{dV_1}{V_1}$$

Integrating and referring to the final and initial conditions gives

$$\Delta S_1 = nRx_1 \ln\left(\frac{V}{V_1}\right)$$

where as usual the symbol x means mole fraction. However

$$\frac{V}{V_1} = \frac{nRT}{p} \times \frac{p}{n_1 RT} = \frac{n}{n_1} = \frac{1}{x_1}$$

according to the initial conditions. Adding in the contribution of the second compartment gives rise to the formula for ideal gas mixing:

$$\Delta S = -nR\left(x_1 \ln x_1 + x_2 \ln x_2\right) > 0$$

(E.1)

Gas mixing results in an entropy increase as expected.

E.2. The supposed paradox.

It supposedly results when we let the two components be identical gases. The mixing formula still apparently holds so that $\Delta S > 0$, but when the partition is removed there is no mixing of different gases, and we expect $\Delta S = 0$.

One resolution of the paradox is an appeal to statistical mechanics and the difference in method of treating distinguishable, identical particles vis-à-vis indistinguishable, identical particles[18]. Another

---

[18] For example, Kondepudi and Prigogine, pp. 156-7.

is to assert that the paradox arises from a paralogism[19]:

It contradicts *the atomistic nature of matter* and is inconsistent with the fact that there is no continuous transition between different kinds of molecules (e.g. the atoms H and He).

However, I think that there is a simpler way to eliminate the paradox based on principles of the continuity or discontinuity of thermodynamic properties. The issues become clearer, I believe, if one finds entropy changes as adiabatic, irreversible transitions between initial and final isentropes. In § E.3-E.5, I use a set of numerical data to argue that the paradox is nothing but a phantom. In § E.6, I show how I got my numerical data, and how by a careful consideration of the continuity vis-à-vis discontinuity issues alluded to above, the paradox is removed using only principles of classical thermodynamics. I summarize and conclude in § E.7.

E.3. Preliminary numerical work.

Consider two ideal monatomic gases, helium and neon. Their standard third law entropies at 298.15 K are given by[20]

---

[19] Sommerfeld, § 13.
[20] http://webbook.nist.gov

46

	He (g)	Ne(g)
S° (J/mol-K)	126.15	146.33

In what follows, we consider sets of states at 298.15 K, $V = 24.79$ L, and $n = 1$ total moles of gases in all cases.

Initial states. First consider a set of initial states, that is, states in which two gases are initially separated by a single partition. In all cases, the two compartments will have volumes $V_1 = 3V/4$ and $V_2 = V/4$. The two gases will be helium and/or neon. The symbol n_1 will be the number of moles of helium in V_1. I define four initial states.

State	I1	I2	I3	I4
n_1	¾	¾	2/3	2/3
x_{He}	1	¾	1	2/3
S (J/K)	126.15	126.52	120.71	127.44

The mole fraction of helium given in the third row is for both compartments. Hence, I1 and I3 have helium in both compartments, whereas I2 and I4 have helium in the first compartment but neon in the second.

In I1 and I2, the initial pressure is $p^0 = 1$ bar in both compartments. This is seen as follows for compartment 1 in I2: noting that $RT/V = 1$ bar/mole, we have that

$$p_1 = \frac{n_1RT}{V_1} = \frac{3}{4}\frac{RT}{3V/4} = p^0 = 1 \text{ bar}$$

with a similar argument following for p_2. However, in I3 and I4, the initial pressures are different in the two compartments as follows for compartment 1 in I3:

$$p_1 = \frac{n_1RT}{V_1} = \frac{2}{3}\frac{RT}{3V/4} = \frac{8}{9}\text{bar}$$

An analogous argument gives $p_2 = 4/3$ bar. Choosing I3 and I4 in this way will help to unravel the Gibbs paradox.

Final states. These are the final states that result when the partition is removed. There are three final states of interest. All three final states are at a pressure of one bar according to Boyle's Law.

State	F1	F2	F3
x_{He}	1	¾	2/3
S (J/K)	126.15	131.20	132.88

A condition to bear in mind. Before proceeding to the analysis of the paradox, one should remember that the application of the mixing formula requires that the initial pressures in the two compartments be equal. Hence, if we begin with the initial states I3 or I4, the mixing formula will not be valid. However, that turns out to be an interesting point in the analysis. Now to the analysis.

E.4. The Gibbs paradox fails to appear.

We consider transitions only between initial and final states of the same chemical composition.

Transition	ΔS (J/K)
I2 → F2	4.68
I1 → F1	0

This is where the paradox was supposed to show up, that when I replaced neon in the second compartment with helium, then because the mixing formula was still valid, I was supposed to get $\Delta S = 4.68 \, J/K$, but evidently that expectation was nothing but a phantom. When both compartments contain helium at the same temperature and pressure, and the partition is removed, the state point remains at the same place on the helium adiabatic curve, as common sense said should be the case all along. Because the two compartments are isobaric, then the mixing equation is valid for the I2 → F2 transition:

$$\Delta S = -R\left[0.75\ln(0.75) + 0.25\ln(0.25)\right] = 4.68 \, J/K$$

E.5. The paradox makes a false appearance.

Consider these transitions:

Transition	ΔS (J/K)
I4 → F3	5.44
I3 → F1	5.44

Now that's what the paradox is supposed to look like—replace neon in compartment 2 with helium, so that the system is all helium, and the system acts as though the heliums in the two compartments were two different chemicals, that Eq. 5.1 was still valid. Of course, that's a red herring, as the mixing equation isn't valid because the initial pressures in the two compartments weren't equal. We see it doesn't work for the I4 → F3 transition, and it *shouldn't* work:

$$\Delta S = -R\left[\frac{2}{3}\ln\left(\frac{2}{3}\right)+\frac{1}{3}\ln\left(\frac{1}{3}\right)\right] = 5.29\,\text{J}/\text{K}$$

which is a meaningless number. The change in entropy arises purely from the smoothing out of pressures—the identities of the gases are irrelevant.

$$\Delta S = -nR\left(x_1\ln\left(\frac{P_{1f}}{P_{1i}}\right)+x_2\ln\left(\frac{P_{2f}}{P_{2i}}\right)\right)$$

by the fundamental equation, where i and f refer to initial and final values. By Boyle's Law

$$\frac{V_{ki}}{V_{kf}} = \frac{P_{kf}}{P_{ki}}$$

where the left-hand side is ¾ for $k = 1$ and ¼ for $k = 2$. Then the entropy change is

$$\Delta S = -R\left[\frac{2}{3}\ln\left(\frac{3}{4}\right)+\frac{1}{3}\ln\left(\frac{1}{4}\right)\right] = 5.44\,\text{J}/\text{K}$$

This was a phony appearance of the so-called Gibbs paradox, because it's an effect of non-isobaric initial conditions, and has nothing to do with the difference in molecular identity.

E.6. Calculation of the entropies of the initial states. The problem is find the entropy of the gas in one compartment. To do that it is convenient to reverse the direction of the mixing problem and start with n_1 moles of a chemical whose standard entropy is $S_1{}^0$. Initially the ideal gas occupies a volume V with an entropy $n_1 S_1{}^0$, but in the final state the volume is V_1. Designate the entropy of the gas in the compartment as S_1. As $dU = TdS - pdV = 0$ by the fundamental equation and the properties of the ideal gas, the process being isothermal, then

$$dS = \frac{pdV}{T} = nR\frac{dV}{V}$$

Integrating for each component, and adding the two entropies together

$$S = S_1 + S_2$$

gives the initial state in the mixing problem:

$$S = n_1\left[S_1^0 + R\ln\left(\frac{V_1}{V}\right)\right] + n_2\left[S_2^0 + R\ln\left(\frac{V_2}{V}\right)\right]$$

For the conditions stated in § E.3, this becomes

$$S = n_1\left[126.15 + R\ln(0.75)\right] + n_2\left[S_2^0 + R\ln(0.25)\right]$$

or

$$S = 123.76n_1 + n_2\left[S_2^0 - 11.53\right]$$

$$(\text{E.2})$$

This is the working equation for calculating the entropy of the initial states, but with this proviso, that it is not applied to I1.

There's a possible objection. If I apply it to I1 I find that the entropy is 121.47 J K^{-1}, and the entropy change in the transition I1 → F1 is

$$\Delta S = 126.15 - 121.47 = 4.68 \, \text{J K}^{-1}$$

But that is the Gibbs paradox—it hasn't gone away, and my claim that it isn't there is in my inconsistent use of Eq. (E.2).

However, that objection doesn't hold: *I have not been inconsistent because Eq. (E.2) isn't valid for state I1*. This is because I1 is qualitatively and quantitatively different from I2, I3, and I4. In I1, the chemical and physical conditions on the two sides of the partition are the same—both sides have the same pressure, the same chemical, and thus the same molar entropy; to get the total entropy I simply add

$$n_1 S_1^0 + n_2 S_1^0 = 126.15 \, \text{J K}^{-1}$$

as the total number of moles is unity. In the above formula I have emphasized that

$$S_2^0 = S_1^0$$

By contrast, there is a chemical discontinuity in I2, a pressure discontinuity in I3, and both in I4.

The whole issue is whether I have correctly calculated the entropy of state I1; if I haven't, then I haven't correctly gotten rid of the paradox. I offer two examples by way of asserting that I am correct.

Suppose that Erlenmeyer flask number one contains 0.750 L of acetone and flask two 0.250 L, both at 298.15 K and 1 bar, for which the molar entropy is $S^0 = 200.4$ J mol^{-1} K^{-1}. From the mass density and molar mass I find that the entropy of acetone in flask one is 2.04 kJ K^{-1} and in two 0.68 kJ K^{-1}. Because the physical conditions of both samples are the same, and because the glass partitioning them does not change their chemical properties, then the total entropy of the pair is 2.72 kJ K^{-1}. If I mix the two chemicals the total entropy is still 2.72 kJ K^{-1}; there is no entropy change. The partitioning has no affect on the state point on the isentropic surface as long as the chemical identity and physical properties are continuous across the glass partition.

Consider a second example. The number of partitions is irrelevant. Given a mole of helium (or acetone) one can insert an "infinite" number of partitions, thereby dividing it into an infinitude of infinitesimal strips. As long as the molar entropy and the pressure are the same in every strip, then

$$S^0 dn$$

is the entropy of a strip at one bar pressure that contains dn moles, and for a system that contains one mole, the entropy is

$$S = \int_0^S dS' = S^0 \int_0^1 dn = S^0$$

As to the entropy value $S = 121.47$ J K^{-1} that seemed to bring back the paradox, it belongs to physical conditions inconsistent with those in the example. For instance, at one bar pressure this is the molar entropy at 238.04 K, meaning that state F1 is inaccessible to it by the *isothermal*, adiabatic, irreversible mixing of ideal gases.

E.7. Summary and conclusions. It will be useful to write the formulas for the initial states in a different form. For convenience, let

$$f(x) = x_1 \ln(x_1) + x_2 \ln(x_2)$$

and

$$g(x,p) = x_1 \ln\left(\frac{p^0}{p_1}\right) + x_2 \ln\left(\frac{p^0}{p_2}\right)$$

The entropies of I2, I3, and I4 by (E.2), given that this equation is not valid for I1, are

$$I1 : S = S_1^0$$

$$I2 : S = x_1 S_1^0 + x_2 S_2^0 + Rf(x) \quad \text{for } x_1 = 3/4$$

$$I3 : S = S_1^0 + R\left[f(x) + g(x,p)\right]$$

$$I4 : S = x_1 S_1^0 + x_2 S_2^0 + R\left[f(x) + g(x,p)\right] \quad \text{for } x_1 \neq 3/4$$

The standard entropies on the right-hand side of each equation are the final states F1, F2, and F3. Hence, the entropy changes are

$$\Delta S = F1 - I1 = 0$$

$$\Delta S = F2 - I2 = -Rf(x)$$

$$\Delta S = F1 - I3 = -R[f(x) + g(x,p)]$$

$$\Delta S = F3 - I4 = -R[f(x) + g(x,p)]$$

The validity of the ideal gas mixing equation, Eq. (E.1), requires that the mixing occurs isobarically, which in turn requires that $V_1/V = x_1$; I chose the left-hand side to be ¾ for my numerical example. This is the case for the I2 → F2 transition.

For those transitions that occur non-isobarically, $V_1/V = ¾ = x_1 P^0/P_1$, meaning that $x_1 \neq$ ¾; I chose $x_1 = 2/3$ in my examples, which fixes the ratio of pressures. This is the case for the last two transitions.

A simple graphical representation would be to graph entropy against mole fraction. Then isentropes would be horizontal lines and non-zero entropy changes would be vertical lines pointing upward from initial to final states.

States I1 and F1 differ only by a partition in the former, but as it does not affect the entropy, then its removal does not result in an entropy change, and there is no Gibbs paradox. By finding entropy changes as adiabatic, irreversible transitions between isentropes, the so-called Gibbs paradox disappears without having to make an appeal outside of classical thermodynamics.

Appendix F. Carathéodory, Kelvin, and Clausius: A Comparison.

F.1. Introduction. Enrico Fermi in his *Thermodynamics* shows that if the Kelvin statement is true, then the Clausius statement must be true, and if it is false, so also for the Clausius statement. This is called establishing the equivalence of the two statements. Establishing the equivalence of the Carathéodory statement with the other two is a disideratum, but given Chandrasekhar's cautionary remark quoted in note 5 above, I am not about to over-reach given his formidable reputation. Hence, I set two goals for this section:

1) To argue that the Carathéodory Second Law strengthens the other two statements; and

2) To present arguments that *suggest* equivalence without claiming to have proved it.

F.2. Kelvin's statement. It is impossible for a system working in a cycle to withdraw heat from a source at a uniform temperature throughout and completely convert it into work with no other changes occurring.

The first two steps of a Carnot cycle meet two of the results specified in the Kelvin statement:

i) Starting at State A, the system withdraws heat $q_1 > 0$ from a source at temperature T_1 by expanding isothermally to State B;

ii) The system converts this completely into work by expanding adiabatically to State C.

$$w_{BC} = -\int_{V_B}^{V_C} p\,dV = -q_1 < 0$$

In making these arguments I make no assumption about the working fluid of the system; in particular, I do not assume that it is an ideal gas.

Now the problem arises as to how the system returns to point A to complete the cycle:

$$\oint dU = 0$$

The usual procedure, as given in Fermi's book, is to appeal to experience with heat engines, which is that they produce waste heat. In the Carnot cycle, compress the system isothermally at T_2 from State C to State D, expelling an amount of heat $q_2 < 0$ from the system.[21] Then one can compress the system adiabatically to State A, completing the cycle[22].

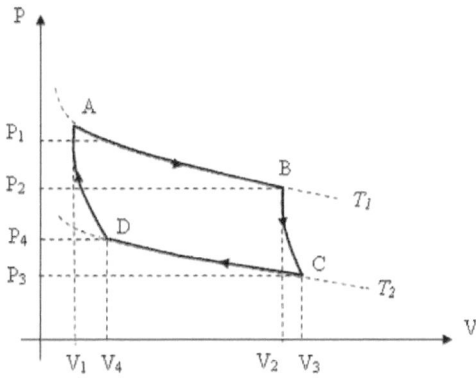

The argument has established that the Kelvin statement is correct because another change has occurred, viz., the expulsion of waste heat from the

---

[21] The sign is necessarily negative. If it were positive, then one could return heat q_1 to its source, with the result that one had removed heat q_2 from a source at a uniform temperature T_2, completely converted it into work, and returned the system to its initial state with no other changes occurring, in violation of the Kelvin statement.

[22] thecarnotengine.blogspot.com Figure 3.

system. But couldn't one wonder if it's possible to have a heat engine that produces no waste heat. Here's where the Carathéodory principle makes the argument stronger: curves BC and DA are adiabats, and the Carathéodory statement says that DA is inaccessible from BC by an adiabatic process; in other words, to reach DA from BC there must be another heat flow.

This is also something like an equivalence argument. If the Carathéodory statement is true, then the Kelvin statement is true. If the Kelvin statement is false, then the Carathéodory statement cannot be correct.

F.3. Clausius' statement. It is impossible for a system working in a cycle to remove heat from a source at a uniform temperature throughout and completely transfer it to a sink at a higher temperature with no other changes occurring.

Let's run the Carnot cycle in the opposite direction to attempt a violation of the Clausius statement:

i) Do isothermal work at T_2 on the system by expanding from State D to State C, which draws an amount of heat $q_2 > 0$ from the source;

ii) Compress the system adiabatically from State C to State B, which raises the temperature to T_1;

iii) Compress the system isothermally at T_1 from State B to State A, which expels an amount of heat $q_1 < 0$ into the high temperature sink;

iv) Expand the system adiabatically from State A to State D, thereby completing the cycle.

We have

$$\oint dU = 0 = \oint dq + \oint dw$$

If

$$\oint dq = q_1 + q_2 = 0$$

then the Clausius statement has been violated. But this would imply that

$$\oint dw = 0$$

However, this cyclic integral evaluated counterclockwise is just the area enclosed by the Carnot cycle. Apparently the two adiabats and two isotherms would have to coalesce to zero area. But the two adiabats cannot coalesce by Carathéodory's statement because they are adiabatically inaccessible; in other words, it prohibits

$$\oint dq = q_1 + q_2 = 0$$

and hence requires an additional heat term, which we add to the above equation.

$$\oint dq = q_1 + q_2 + q \neq 0$$

This saves the Clausius statement because something else has occurred, viz., an additional heat flow.

What is the nature of this added heat term? Let's assume that the original aim of the cycle is accomplished, that $q_1 + q_2 = 0$. Then for the cycle

$$\oint dw = -\oint dq = -q > 0$$

The cyclical work term is positive when the cycle is taken in the counterclockwise direction. Hence, q < 0, which means that some work has been degraded: it has produced a waste energy that is expelled from the system

In conclusion, it seems that in order for the Clausius statement to be true, so must the Carathéodory. If the former is false, the latter cannot be true.

Appendix G. A vector representation and the necessary condition of the Mathematical Theorem.

G.1. Vector representation. Let a generic force be represented by

$$\vec{F} = \hat{i}X + \hat{j}Y + \hat{k}Z$$

(G.1)

and a line element by

$$d\vec{r} = \hat{i}dx + \hat{j}dy + \hat{k}dz$$

(G.2)

Then a generic energy is given by

$$dq = \vec{F} \cdot d\vec{r} = Xdx + Ydy + Zdz$$

(G.3)

Here are the issues: whether dq is exact; if it is not, whether an integrating multiplier exists; and if one exists, then the issue of interest is the nature of the geometric space associated with the condition that dq = 0. The last-named issue means that we are interested in those instances where the force and line element are orthogonal. The vector formulation is useful for addressing these issues. In this appendix, I will not be addressing thermodynamic problems per se but I give a simple example of how a thermodynamic problem can be formulated vectorially.

I consider a bivariant example. By the First Law, one can show that for a reversible adiabatic change of state in an ideal gas,

$$\frac{dT}{T} + (\gamma - 1)\frac{dV}{V} = 0 \qquad (G.4)$$

Let

$$d\vec{r} = \hat{t}\,dT + \hat{v}\,dV$$

and the generic force be

$$\vec{F} = \frac{\hat{t}}{T} + (\gamma - 1)\frac{\hat{v}}{V} \qquad (G.5)$$

where I postulate that the unit vectors are orthogonal. Then

$$dq = \vec{F} \cdot d\vec{r} = 0$$

recovers Eq. (G.4). As dq is exact, the force can be replaced by the gradient of a potential, and dq by the potential's differential: the preceding equation becomes

$$d\xi = \nabla\xi \cdot d\vec{r} = \frac{\partial\xi}{\partial T}dT + \frac{\partial\xi}{dV}dV = 0$$

The partial derivatives of course are given by the coefficients of the unit vectors in Eq. (G.5), so

$$d\xi = d\ln\left(TV^{\gamma-1}\right) = 0$$

leading to the familiar result by a less-traveled road

$$TV^{\gamma-1} = \Gamma = \exp(\xi)$$

(G.6)

which is a family of non-intersecting curves in the T,V-plane each characterized by a different value of Γ. To repeat a familiar mantra, accessible state points exist on a curve given by a particular value of Γ, and inaccessible neighboring points on curves given by $\Gamma \pm d\Gamma$.

G.2. On the existence of integrating multipliers.

For the generic force given by Eq. (G.1), construct the following vector formula:

$$\vec{F} \cdot \nabla \times \vec{F} = C$$

(G.7a)

Expanding gives

$$X\left(\frac{\partial Z}{\partial y} - \frac{\partial Y}{\partial z}\right) + Y\left(\frac{\partial X}{\partial z} - \frac{\partial Z}{\partial x}\right) + Z\left(\frac{\partial Y}{\partial x} - \frac{\partial X}{\partial y}\right) = C$$

(G.7b)

Equations (G.7) have the following remarkable properties:

1) If dq is exact, $C = 0$. This follows immediately as curl(F) is identically zero.

2) If dq is inexact and $C = 0$, then an integrating multiplier exists. As an example consider Eq. (2.6), where the left-hand side of Eq. (G.7b) gives

$$\frac{1}{xyz}\left(\frac{1}{x^2}-\frac{1}{x^2}+\frac{1}{y^2}-\frac{1}{y^2}+\frac{1}{z^2}-\frac{1}{z^2}\right)=0$$

3) If dq is inexact and C ≠ 0, then an integrating multiplier does not exist. As an example, the Pfaffian in Appendix A gives

$$X(0-0)+Y(0-0)+Z(-1-1)=-2z$$

It is useful to note that for a bivariant system, C = 0 always, that is, a bivariant Pfaffian is always integrable. It is easy to prove this by evaluating curl(F) for a bivariant system and showing that the vector F is always orthogonal to it. Evidently systems that are trivariant or higher may not be integrable.

G.3. Proof of the properties of Eqs. (G.7).

I assume case 2) above, that dq is inexact, but that an integrating multiplier exists, and show that C = 0.[23] As dq given by Eq. (G.3) is assumed to be inexact, and since the assumption is that there is an integrating multiplier, then

$$dG = \lambda\,dq$$

will be exact. This is equivalent to having a generic force

$$\vec{G}(x,y,z)=\lambda(x,y,z)\vec{F}(x,y,z)$$

As dG is exact, then the generic force can be equated to the gradient of a potential:

$$\vec{G}=\nabla\left(\xi\right)$$

---

[23] Kestin, *A Course in Thermodynamics*, § 10.2.2.

This is always permitted as the curl of a gradient is always identically zero:

$$\nabla \times \vec{G} = \nabla \times \nabla(\xi) \equiv 0$$

Let

$$\vec{G} = \hat{i}P + \hat{j}Q + \hat{k}R$$

where $P = \lambda X$, $Q = \lambda Y$, and $R = \lambda Z$. Then

$$\nabla \times \vec{G} = \hat{i}\left(\frac{\partial R}{\partial y} - \frac{\partial Q}{\partial z}\right) + \hat{j}\left(\frac{\partial P}{\partial z} - \frac{\partial R}{\partial x}\right) + \hat{k}\left(\frac{\partial Q}{\partial x} - \frac{\partial P}{\partial y}\right) = 0$$

Each term in parentheses must vanish separately. Now multiply the first bracket by X, the second by Y, and the third by Z, and then evaluate the derivatives. It will suffice to show one example:

$$X\left(\frac{\partial R}{\partial y} - \frac{\partial Q}{\partial z}\right) = \lambda X\left(\frac{\partial Z}{\partial y} - \frac{\partial Y}{\partial z}\right) + XZ\frac{\partial \lambda}{\partial y} - XY\frac{\partial \lambda}{\partial z} = 0$$

Do the same for the other two parenthetical expressions, add the three together, and the terms involving the derivatives of the integrating multiplier will vanish, leaving only Eq. (9.7) with C = 0, which was the proposal to be proved.

G.4. That the Mathematical Theorem of § 2 expresses a necessary condition.

Given the Pfaffian differential Eq. (G.3), then Eq. (G.7) with C = 0 means that dq is integrable, that it is either exact or can be made exact because an integrating multiplier exists. Then the adiabatic condition that dq = 0 gives rise to a family of surfaces, $\xi(x,y,z) = \Gamma$. This in turn means that in the vicinity of any state point H_0 there will be states H that are inaccessible from the surface containing H_0.

64

Hence, the condition dq = 0, the existence of integrating mulitipliers, and the condition of inaccessibility are linked together by mathematical necessity. Thus in Carathéodory's statement of the Second Law, not only is the axiom of inaccessibility a sufficient condition for the existence of an integrating multiplier, but a necessary one.

Appendix H. A spectrum of viewpoints on the Carathéodory second law.

Joseph Kestin writes[24] that

C. Carathéodory…succeeded in providing an alternative formulation [of the Second Law]…In this manner he was able to achieve greater elegance and generality of exposition. The only obstacle in appreciating this economical formulation is presented by the general lack of familiarity with the mathematical propositions required for the purpose.

Of course my purpose has been to write clearly about the required mathematical propositions.

The monograph by Margenau and Murphy has been a good companion for more than fifty years. The opening chapter of thirty-one pages is "The Mathematics of Thermodynamics", including a

---

[24] Kestin, *A Course in Thermodynamics*, p. 457.

thorough treatment of Pfaffians and integrating multipliers. The last section of the first chapter is "The Principle of Carathéodory", which they do not treat in detail. Born notes,[25] however, that they have conflated the Mathematical Theorem (§ 2 above) and Carathéodory's statement of the second law (§ 4).

I still own the book in which I bit into the "entropic apple" for the first time, viz., Irving Klotz's textbook, now nearly seventy years old. He calls the Carathéodory approach the "most formal and elegant approach to thermodynamics,"[26] but notes that chemistry students usually have little acquaintance with the Pfaffian equations on which it is based. It is probably for that reason that he doesn't use it in developing thermodynamic theory and applications. Pfaffians are, of course, the point of departure for this mini-monograph.

The monograph *Chemical Thermodynamics* by Lewis and Randall, as revised by Pitzer and Brewer, which is something of a sacred text among physical chemists, seems not to mention Carathéodory at all.

A.B. Pippard finds that Carathéodory's second law hasn't the direct appeal to the engineer

---

[25] Born, *Natural Philosophy*, p. 143.
[26] Klotz, *Chemical Thermodynamics*, p. 4.

or the "practically minded physicist" that's found in the Kelvin and Clausius statements, and that "it is neither intuitively obvious nor supported by a mass of experimental evidence. It may be argued therefore that the further development of thermodynamics should not be made to rest on this basis,..."[27]

C. Truesdell in his *Rational Thermodynamics* roundly condemns[28] the Carathéodory approach, in effect repudiating point-by-point what Born finds efficacious in it. There is no point, he says, in redoing the proof of a problem that has already been solved informally. It is "axiomatization for axiomatization's sake..." Its biggest "disaster" is attempting to define heat in mechanical terms as "[a]ny such attempt must fail."

J. R. Partington in his massive treatise says of Carathéodory's approach, "The method is too abstract to appeal to many students of physical chemistry."[29]

Berry et al.[30] in their equally massive treatise give equal status to all three statements of the Second Law. They ask, "are there thermodynamic

[27] Pippard, *Elements of Classical Thermodynamics*, pp. 30-31.

[28] Quoted by Georgiadou on p. 50 of her biography of Carathéodory. I have not seen Truesdell's book.

[29] *Advanced Treatise*, p. 178.

[30] *Physical Chemistry*, Chapter 16.

states of the system that are inaccessible by use of the mechanical processes that define U? The answer to this comes from experiment." The answer is yes. There is a short discussion of Pfaffian differentials, which they aptly call linear differentials, and of integrating multipliers.

Appendix I. Biographical sketch of Constantin Carathéodory.

"Constantin Carathéodory was a cosmopolitan of Greek origin, whose intellectual profile was moulded at the cross roads of European and Oriental culture."[31] Constantin Carathéodory was born in Berlin September 13 1873 and died in Munich February 2 1950. He lived most of his life in Germany and Belgium, and with his family lived in Germany through the privations of both world wars.

At the time of his birth his father Stephanos was a diplomat, the Ottoman ambassador in Brussels, a post not to be confused with that of the Turkish ambassador to Belgium. The Carathéodorys were well-connected with the Ottoman government in Constantinople. Constantin took his education in Brussels, graduating in 1896 as an engineer from the Military

---

[31] The opening sentence of Georgiadou's big biography. It's too classy not to quote. My biographical sketch derives entirely from her book.

School of Belgium. For about four years he earned his living through engineering projects in Greece and Egypt.

In 1900 he turned to the study of mathematics, earning the degree of D. Phil. in 1904 from the University of Göttingen with a dissertation *On the Discontinuous Solutions of the Calculus of Variations.* For a number of years he lived variously at Bonn, Hannover, Göttingen, and Constantinople. He was in the last-named city from July-September 1908 where he worked out his theory of thermo-dynamics, a subject that had attracted him during his student days in Brussels. He said of his 1908 work that "it was the only work that he had thought through from the beginning to the end before he wrote it down."[32]

He joined the mathematics faculty as a professor at the University of Göttingen in 1913, where he was awarded the chair held by the renowned Felix Klein, who had retired. He remained at Göttingen until 1918.

Following the First World War, the Greek Prime Minister Eleutherios Venizelos invited him to Greece to found a new university, an invitation that he accepted, and one that would ultimately ensnare him in the great national tragedy of Greece. A bit of history is appropriate.

---

[32] Ibid., p. 47.

The defeat of the Central Powers with their Ottoman ally led to the ceding of territory in Anatolia, that is, in what is now modern Turkey, to Greece. Then, early in 1921, the Greeks launched a military campaign against the Turks in Anatolia. This was the time in which the location of the university was to be decided. Carathéodory proved to be the decisive player in this discussion, and his choice was the city of Smyrna, now the Turkish city Izmir, on the western coast of Anatolia. He was named Rector of the Ionian[33] University of Smyrna, and, essentially the president of the new university, set about hiring faculty, raising money, and planning its campus.

The Turkish army under Mustafa Kemal routed the Greek army in 1922, drove it from Anatolia, and seized the city of Smyrna. The non-Turkish population was plundered, murdered, and raped; a fire lasting three days destroyed the city.[34] Carathéodory was lucky to escape the calamity aboard a Greek battleship.

The Anatolian Disaster subsequently included the expulsion of Greeks from Turkey and of Turks from Greece. About a million-and-a-half Greeks

---

[33] Why "Ionian"? Ionia was a region of ancient Greece in what is now southwest Turkey. More than one historical account recognizes the citizens of the Greek city-state of Miletus in Ionia as the founders of western science. The most famous Milesians are Thales, Anaximander, and Anaximenes.

[34] Georgiadou, p. 167.

suddenly appeared in mainland Greece as refugees, and Carathéodory, instead of founding an institution of higher learning, found himself as one of the leaders in mitigating the sufferings of those displaced from the land that had been home to their ancestors for three thousand years.

In 1924, Carathéodory returned to Germany where he gained a professorship in mathematics at the University of Munich, where he remained until his retirement in 1938.

The Fields Medal is the mathematician's equivalent of the Nobel Prize. Carathéodory never won this medal, but he was chairman of the Fields Medal Committee that made the first awards in Oslo in 1936, which, I think, is a measure of his stature among his peers.[35]

A large section[36] of Georgiadou's biography recounts the life of a humane man living under the inhumane racial laws of the Third Reich. He and his wife, in order to have some measure of security, had to establish that they were Aryan, meaning, according to the Nuremberg Laws, that no parent or grand-parent was non-Aryan (Jewish) and that they did not practice Judaism. As a University professor, he had to declare his loyalty to Hitler.[37] Georgiadou lists name-after-name of scientists and

---

[35] Ibid., p. 317.
[36] Ibid.; mainly Chapter Five
[37] Ibid., p. 469.

mathematicians who were ousted or resigned from universities and scholarly organizations because they were Jewish and/or dissidents, only to be replaced by Nazi loyalists of dubious competence. If they were lucky or a genius like Born they could emigrate. Others were forced into exile and penury or imprisoned or sent to the camps. Some committed suicide. Carathéodory sent his daughter Despina from Munich to Athens for her law studies, because the quality of education at Munich was compromised, and because the Nazis initially discouraged and then prohibited women from practicing law.

Because he remained in Germany, in effect trapped there by the war, and given his international reputation, the Reich could set him forward to the world as an example of the high level of scholarship in Germany. At the same time, the secret police compiled dossiers on him, some highly suspicious that he was too friendly to Jews, that he was dangerous to the Reich, and others that he was loyal to it. Georgiadou's study of these dossiers shows the police's incompetence at least in the case of Carathéodory as they blundered in trying to establish simple facts about his life. There are at least two incidents where he tried to intervene on behalf of mathematicians imprisoned by the Gestapo or SS; in one instance, the man was released from custody, and in the other the man was shot, but it's not clear that Carathéodory had any leverage in the former instance, and clear that it was insufficient in the latter.

The following titles reveal Carathéodory's mathematical (and physical) interests.

Lectures on Real Functions, 1918, revised 1927, 1948.

Conformal Representation, 1932.

The Calculus of Variations and Partial Differential Equations of the First Order, 1935.

Geometric Optics. 1937.

Real Functions: Volume I; Numbers , Point Sets, Functions. 1939, revised 1946. He wrote a second volume for this work, which reached the printers, but was destroyed there in an air-raid in 1943.

Given that much of his mathematics comprises the tool-kit of mathematical physicists, his interactions with physical scientists come as no surprise. These include, of course, Born, but also Einstein, Bohr, Planck, Pauli, Zermelo, and others.

Bibliography

Berry, R. Stephen, Stuart A. Rice, and John Ross. *Physical Chemistry*. Second edition. Oxford University Press, 2000.

Born, Max. *Natural Philosophy of Cause and Chance*. Dover, New York, 1964.

Buchdahl, H. A. "On the Principle of Carathéodory. *American Journal of Physics*, 1949 (1), 41-43. "On the Theorem of Carathéodory". Ibid., 44-46.

Carathéodory, Constantin. "Investigation into the Foundations of Thermodynamics." In *The Second Law of Thermodynamics*. Ed. Joseph Kestin. Dowden, Hutchinson, and Ross, Stroudsberg, PA., 1976. Trans. by Kestin from "Untersuchungen über die Grundlagen Der Thermodynamik", *Math. Ann. (Berlin)* **1909**, *67*, 355-386.

Chandrasekhar, S. *An Introduction to the Study of Stellar Structure*. Dover, New York, 1957. (Reprint of the 1939 edition.)

Fermi, Enrico. *Thermodynamics*. Dover, New York, 1957. (Reprint of the 1939 edition.)

Georgiadou, Maria. *Constantin Carathéodory: Mathematics and Politics in Turbulent Times*. Springer, Berlin, 2000.

Kestin, Joseph. *A Course in Thermodynamics*. Blaisdell, Waltham, Massachusetts. 1966.

Klotz, I.M. *Chemical Thermodynamics*. Prentice-Hall, New York, 1950.

Kondepudi, Dilip, and Ilya Prigogine. *Modern Thermodynamics: From Heat Engines to Dissipative Structures*. John Wiley and Sons, New York, 1998.

Lewis, G. N. and M. Randall. *Chemical Thermodynamics*. Rev. by K. Pitzer and L. Brewer. McGraw-Hill, New York, 1961.

Margenau, Henry and George Murphy. *The Mathematics of Physics and Chemistry*. Second edition. Van Nostrand, Princeton, N.J., 1956.

Partington, J. R. *An Advanced Treatise on Physical Chemistry: Fundamental Principles/The Properties of Gases*. Volume I. Longmans, Green, and Co., London, 1949.

Pippard, A. B. *The Elements of Classical Thermodynamics*. Cambridge University Press, Cambridge, 1966.

Sommerfeld, Arnold. *Thermodynamics and Statistical Mechanics*. In *Lectures on Theoretical Physics, Volume V*. Academic Press, New York, 1964.

www.ingramcontent.com/pod-product-compliance
Lightning Source LLC
Chambersburg PA
CBHW022125170526
45157CB00004B/1749